DUNCAN MACMILLAN (Writer)
Duncan is a former Writer-in-Residence at Paines Plough and Manchester Royal Exchange. Plays include the acclaimed adaptation of George Orwell's 1984, co-adapted/co-directed with Robert Icke (Headlong / Nottingham Playhouse, UK tour, Almeida Theatre and West End); EVERY BRILLIANT THING (Paines Plough / Pentabus); 2071 co-written with Chris Rapley (Royal Court / Hamburg Schauspielhaus); REISE DURCH DIE NACHT adapt. Friederike Mayröcker created with Katie Mitchell and Lyndsey Turner (Schauspielhaus Köln, Theatertreffen, Festival d'Avignon); WUNSCHLOSES UNGLÜCK adapt. Peter Handke (Burgtheater Vienna); THE FORBIDDEN ZONE (Salzburg Festival / Schaubühne Berlin); LUNGS (Studio Theatre Washington DC / Paines Plough & Sheffield Theatres); ATMEN (Schaubühne Berlin); DON JUAN COMES BACK FROM THE WAR adapt. Ödön von Horváth (Finborough Theatre); MONSTER (Royal Exchange / Manchester International Festival). Duncan was the recipient of two awards in the inaugural Bruntwood Playwriting Competition, 2006. Other awards include Best New Play at the Off West End Awards 2013 for LUNGS; the Big Ambition Award, Old Vic 2009; the Pearson Prize, 2008. His work with director Katie Mitchell has been selected for Theatertreffen and the Avignon Festival.
Duncan won Best Director (with Robert Icke) at the UK Theatre Awards for 1984, which was also nominated as Best New Play at the Olivier Awards 2014.

SIAN REESE-WILLIAMS (W)
Theatre credits include: LUNGS, THE INITIATE, OUR TEACHER'S A TROLL (Paines Plough Roundabout Season), ENJOY (West Yorkshire Playhouse), CHILDREN OF FATE (Bussey Building), BE MY BABY (New Vic Theatre), AS YOU LIKE IT (Derby Playhouse), COLTAN (Paines Plough), SIXTY FIVE MILES (Paines Plough/Hull Truck), DIAMOND (King's Head Theatre), THE DREAMING (National Youth Music Theatre), INTO THE WOODS (National Youth Music Theatre).
TV credits include: EMMERDALE (ITV) and COWBOIS AC INJIANS (Opus TF).

ABDUL SALIS (M)
Theatre credits include: THE RISE AND SHINE OF COMRADE FIASCO (Gate Theatre), THE INITIATE, OUR TEACHER'S A TROLL, LUNGS (Paines Plough Roundabout Season), JOE GUY (Tiata Fahodzi), EXONNERATED, WAR HORSE (National Theatre), DON JUAN IN SOHO (Donmar Warehouse) and HENRY V (Unicorn).
TV credits include: DOCTORS (Blunt Pictures), HACKS (Hat Trick Productions), STRIKE BACK (Left Bank Pictures), OUTNUMBERED (Hat Trick Productions), VICTORIA WOOD CHRISTMAS SPECIAL (Phil McIntyre Productions), CASUALTY (BBC). M.I. HIGH (Kudos for BBC), THE BILL (Talkback Thames), DOCTOR WHO (BBC/DW Productions), GIFTED (Red Productions), TREVOR'S WORLD OF SPORT (Hat Trick Productions), ROGER ROGER (BBC), THE HIDDEN

CITY (Hallmark Entertainment).
Film credits include: FLY BOYS (Electric Entertainments), ANIMAL (Animal Productions), SAHARA (Sahara Productions), WELCOME HOME (Wega-Film), LOVE, ACTUALLY (DNA/Working Title).
Radio includes: SKYVERS (BBC Radio 3).

GEORGE PERRIN (Direction)

George Perrin is the joint Artistic Director of Paines Plough. He was formerly co-founder and Joint Artistic Director of nabokov and Trainee Associate Director at Paines Plough and Watford Palace Theatre.

Directing credits for Paines Plough include LUNGS by Duncan Macmillan, THE INITIATE by Alexandra Wood, OUR TEACHER'S A TROLL by Dennis Kelly (Roundabout Season, Edinburgh Festival Fringe and National Tour), NOT THE WORST PLACE by Sam Burns (Sherman Cymru, Theatr Clwyd), SEA WALL by Simon Stephens (Dublin Theatre Festival / National Theatre Shed), GOOD WITH PEOPLE by David Harrower (59East59 Theatres New York / Traverse Theatre / Oran Mor), LONDON by Simon Stephens (National Tour), SIXTY FIVE MILES by Matt Hartley (Hull Truck), THE 8TH by Che Walker and Paul Heaton (Latitude Festival/ Barbican / Manchester International Festival / National Tour), DIG by Katie Douglas (Oran Mor / National Tour) and JUICY FRUITS by Leo Butler (Oran Mor / National Tour).

As Trainee Associate Director of Paines Plough, directing credits include HOUSE OF AGNES by Levi David Addai, THE DIRT UNDER THE CARPET by Rona Munro, CRAZY LOVE by Che Walker, MY LITTLE HEART DROPPED IN COFFEE by Duncan Macmillan and BABIES by Katie Douglas.

Further directing credits include 2ND MAY 1997 by Jack Thorne (Bush Theatre), TERRE HAUTE by Edmund White (59East59 Theatres New York, West End, National Tour and Assembly Rooms, Edinburgh Festival Fringe), IS EVERYONE OK? and PUBLIC DISPLAYS OF AFFECTION by Joel Horwood and CAMARILLA by Van Badham (nabokov).

LUCY OSBORNE (Designer)

Lucy's recent theatre credits include THE ANGRY BRIGADE (Paines Plough / Plymouth Theatre Royal), LAMPEDUSA (Soho Theatre/Hightide Festival), HELLO GOODBYE and IN THE VALE OF HEALTH (Hampstead Theatre), PRIVACY, CORIOLANUS and THE RECRUITING OFFICER (Donmar Warehouse), THE ROUNDABOUT SEASON (PainesPlough), AN INTERVENTION (Paines Plough/Watford Palace Theatre), TRANSLATIONS (Sheffield Crucible/ETT/Rose Kingston/ UK Tour, winner 'Best Touring Production' at the UK Theatre Awards), THE MACHINE (Manchester International Festival/ The Amoury, New York).

Lucy is a former Associate Artist at the Bush Theatre and designed JUMPERS FOR GOALPOSTS, ARTEFACTS, WRECKS, THE BROKEN SPACE SEASON, 2,000 FEET AWAY, TINDERBOX and tHe dYsFUnCKshOnalZ!.

Other theatre credits include

TWELFTH NIGHT (Chicago Shakespeare Theatre, winner of the Jeff Award for Best Design), PLENTY (Sheffield Theatres), CLOSER (Theatre Royal Northampton), THE LONG AND THE SHORT AND THE TALL (Sheffield Lyceum), THE UNTHINKABLE (Sheffield Crucible Studio).
Lucy is a co-designer of Roundabout, a pop-up theatre in the round which won The Stage Awards 'Theatre Building of the Year 2015'.

EMMA CHAPMAN
(Lighting Designer)
Emma Chapman trained at Bristol Old Vic Theatre School.
With Lucy Osborne and Howard Eaton, Emma Chapman conceived and realised The Stage Awards' *Theatre Building of the Year 2015*: Roundabout, commissioned by Paines Plough. That company took three plays on tour in this pop up venue, including the Edinburgh Festival 2014. Roundabout is currently on its 2015 UK Tour.
Theatre credits include: a co-production between The Leeds Theatre Trust and West Yorkshire Playhouse of BOI BOI IS DEAD; four shows including Rose with Art Malik at the 2013 Edinburgh Fringe, THE PLANET and STUFF AND RUN for Polka Theatre; two plays for Bath Theatre Royal Ustinov Studio; DUBLIN CAROL (Donmar season) and SEX WITH A STRANGER with Russell Tovey and Jaime Winston at Trafalgar Studios; the acclaimed production of O'Neill's THREE SEA PLAYS in the Old Vic TUNNELS; PARALLEL HAMLET at the Young Vic; DICK WHITTINGTON in Bury St Edmunds.
Opera credits include XERXES and CARMEN (Royal Northern College of Music, Manchester), COSÌ FAN TUTTE (Royal College of Music), THE PIED PIPER (Opera North) and Il Turco in ITALIA (Angers/Nantes Opera an Luxembourg).
She has also lit RUMPLESTILTSKIN for London Children's Ballet at the Peacock Theatre, available on DVD.
Other notable engagements include Olivier Award-winning play THE MOUNTAINTOP for Theatre 503 and at Trafalgar Studio; the highly praised THE PAINTER which opened the new Arcola Theatre, and the David Mamet double bill which opened their studio; WET WEATHER COVER at the King's Head and Arts Theatres; THE MACHINE GUNNERS (Polka Theatre).

TOM GIBBONS
(Sound Designer)
Tom trained at Central School of Speech and Drama.
Theatre Includes: ELEPHANTS (Hampstead Theatre), HENRY IV (Donmar Warehouse), THE ANGRY BRIGADE (Paines Plough), BREEDERS (St. James Theatre), Roundabout Season 2014 (Paines Plough), THE WHITE DEVIL (RSC), MR BURNS (Almeida Theatre), A VIEW FROM THE BRIDGE (Young Vic/Wyndham's – nominated for Best Sound Design Olivier 2015), HAPPY DAYS (Young Vic), AN INTERVENTION (Paines Plough/Watford Palace Theatre), HOPELESSLY DEVOTED (Paines

Plough), TRANSLATIONS (Sheffield Crucible), HOME (Arcola), 1984 (Headlong/Almeida/West End), LIONBOY (Complicite), GROUNDED (Gate Theatre/Traverse Theatre), AS YOU LIKE IT (RSC), JULIUS CAESAR (Donmar), HITCHCOCK BLONDE (Hull Truck), THE SPIRE (Salisbury Playhouse), LONDON (Paines Plough), ROUNDABOUT SEASON 2013 (Shoreditch Town Hall, Paines Plough), THE ROVER (Hampton Court Palace), LOVE LOVE LOVE, (Royal Court), ISLAND (National Theatre, Tour), ROMEO & JULIET (Headlong), DISCO PIGS (Young Vic), DEAD HEAVY FANTASTIC (Liverpool Everyman), PLENTY (Crucible Studio, Sheffield), ENCOURAGE THE OTHERS (Almeida), WASTED (Paines Plough, Tour), CHALET LINES, THE KNOWLEDGE, LITTLE PLATOONS, 50 WAYS TO LEAVE YOUR LOVER, (Bush Theatre) HAIRY APE, SHIVERED, FAITH, HOPE AND CHARITY, THE HOSTAGE, TOAD (Southwark Playhouse), SOLD (503), THE CHAIRS (Ustinov Bath), THE COUNTRY, THE ROAD TO MECCA, THE ROMAN BATH, 1936, THE SHAWL (Arcola), UTOPIA, BAGPUSS, EVERYTHING MUST GO, SOHO STREETS (Soho Theatre), THE MACHINE GUNNERS (Polka), FAT (The Oval House, Tour), JUST ME BELL (Graeae, Tour), FANTA ORANGE, BLUE HEAVEN (Finborough), RHINEGOLD (The Yard)
As Associate: A SEASON IN THE CONGO (Young Vic), CHOIR BOY (Royal Court), BROKEN SPACE SEASON (Bush Theatre), ORESTEIA (Almeida), ANNA KARENINA (Manchester Royal Exchange) and THE ABSENCE OF WAR (Headlong).

KATE SAGOVSKY (Movement Director)
Kate trained in Dance Studies at Laban after completing an MA in Movement Studies at The Royal Central School of Speech & Drama, & a degree in English Literature at Oxford University. She works across theatre, dance and live art as a director and movement director/choreographer.
Kate has worked widely as a movement director, including on productions at Theatre Royal Stratford East, BAC, the Bush, Oxford Playhouse, Tristan Bates and Theatre503. She worked as the Resident Movement Practitioner at the Royal Shakespeare Company for the 2011 Season, including movement direction on THE HOMECOMING (dir. David Farr) and MOJO (dir. Justin Audibert). She continues to work for the RSC as a Freelance Movement Practitioner and Education Associate Practitioner. She has also taught as a lecturer in Actor-Movement at many UK Drama Schools, at AFDA Film School (Cape Town, South Africa), and for Shakespeare's Globe.
As Artistic Director of MOVING DUST Kate creates cross-art-form performance that has toured to theatre & dance festivals around the UK [www.movingdust.com]. Other productions as Director include A MIDSUMMER NIGHT'S DREAM (Cambridge Arts Theatre), & LOVE'S LABOUR'S LOST (The Metropolitan Arts Centre, Tokyo & UK Tour).

Work as Associate Director (Movement) includes: ESKA'S ENGLISH SKIES (Queen Elizabeth Hall, Southbank Centre) and GVE with Matthew Herbert Big Band (Barbican/Glastonbury). She has also worked as a Staff Director at the National Theatre.

SEAN LINNEN
(Associate Director)
In 2014, Sean was Trainee Artistic Director at Paines Plough and Sheffield Theatres in partnership with Arts Council England. Directing credits include: GROWTH by Luke Norris (Paines Plough/RWCMD); A PREOCCUPATION WITH ROMANCE by Beth Grant (Edinburgh Festival Fringe); HOLLOW by Beth Grant (Bike Shed Theatre, Exeter).
As Associate/Staff Director: PROTEST SONG by Tim Price (National Theatre; LUNGS by Duncan Macmillan; OUR TEACHER'S A TROLL by Dennis Kelly and THE INITIATE by Alexandra Wood (Paines Plough). Other assistant director credits include: TRANSLATIONS by Brian Friel (ETT/Rose Theatre, Kingston/Sheffield Theatres); THE MACHINE by Matt Charman (Donmar Warehouse/ Manchester International Festival/ Park Avenue Armory, New York); THE DAUGHTER-IN-LAW by D.H.Lawrence; A TASTE OF HONEY by Shelagh Delaney and COPENHAGEN by Michael Frayn (Sheffield Theatres).

DOMINIC KENNEDY
(Associate Sound)
Dominic Kennedy is a Sound Designer and Composer for performance and live events; he has a keen interest in developing new work and implementing Sound Design at an early stage of a process. Dominic is a graduate from RCSSD where he developed specialist skills in collaborative and devised theatre-making. Dominic has worked as an Associate Sound Designer with Tom Gibbons on a number Paines Plough productions including LUNGS, THE INITIATE, HOPELESSLY DEVOTED and THE ANGRY BRIGADE, he also composed original music for OUR TEACHER'S A TROLL. Recent Sound Design credits include CROCODILES at the Manchester Royal Exchange, ONO at Oval House, THIS IS THE MOON at The Yard, RUN at the New Diorama.

"Revered touring company Paines Plough" **Time Out**

Paines Plough is the UK's national theatre of new plays. We commission and produce the best playwrights and tour their plays far and wide. Whether you're in Liverpool or Lyme Regis, Scarborough or Southampton, a Paines Plough show is coming to a theatre near you soon.

"The lifeblood of the UK's theatre ecosystem" **The Guardian**

Paines Plough was formed in 1974 over a pint of Paines bitter in the Plough pub. Since then we've produced more than 150 new productions by world renowned playwrights like Stephen Jeffreys, Abi Morgan, Sarah Kane, Mark Ravenhill, Dennis Kelly and Mike Bartlett. We've toured those plays to hundreds of places from Manchester to Moscow to Maidenhead.

"That noble company Paines Plough, de facto national theatre of new writing" **The Daily Telegraph**

Our Programme 2015 sees 11 productions by the nation's finest writers touring to 74 places from Cornwall to the Orkney Islands; in village halls and Off-Broadway, at music festivals and student unions, online and on radio, and in our own pop-up theatre Roundabout.

"I think some theatre just saved my life"
@kate_clement on Twitter

Paines Plough Limited is a company limited by guarantee and a registered charity.
Registered Company no: 1165130
Registered Charity no: 267523

Paines Plough, 4th Floor, 43 Aldwych, London WC2B 4DN
+ 44 (0) 20 7240 4533
office@painesplough.com
www.painesplough.com

Follow @PainesPlough on Twitter
Like Paines Plough at facebook.com/PainesPloughHQ
Donate to Paines Plough at justgiving.com/PainesPlough

Paines Plough are:

Joint Artistic Directors	James Grieve
	George Perrin
Producer	Hanna Streeter
General Manager	Aysha Powell
Assistant Producer	Francesca Moody
Administrator	Natalie Adams
Trainee Producer	Rachel D'Arcy
Trainee Administrator	Bhavini Goyate
Associate Director	Stef O'Driscoll
Technical Manager	Colin Everitt
Press Representative	Kate Morley
Big Room Playwright Fellowship	Nathan Bryon

Roundabout is Paines Plough's beautiful new pop-up in-the-round theatre. It is a completely self-contained 168-seat auditorium that flat packs into a single lorry and can pop up anywhere from theatres to school halls, sports centres, warehouses, car parks and fields.

We created Roundabout because we're passionate about new plays and we want as many people as possible to be able to see them. Roundabout means we can tour further and wider than ever before. For the next decade Roundabout will travel the length and breadth of the UK bringing the nation's best playwrights and a thrilling theatrical experience to audiences everywhere.

Designed by Lucy Osborne and Emma Chapman in collaboration with Charcoalblue and Howard Eaton.

'A beautifully designed masterpiece in engineering.' (*The Stage*)

Theatre Building of the Year – The Stage Awards 2015

LUNGS

Duncan Macmillan

LUNGS

OBERON BOOKS
LONDON

WWW.OBERONBOOKS.COM

First published in 2011 by Oberon Books Ltd
521 Caledonian Road, London N7 9RH
Tel: +44 (0) 20 7607 3637 / Fax: +44 (0) 20 7607 3629
e-mail: info@oberonbooks.com
www.oberonbooks.com

Reprinted with revisions in 2012 (twice), 2013 (twice), 2014, 2015 (thrice)

A catalogue record for this book is available from the British Library.

PB ISBN: 978-1-84943-145-3
E ISBN: 978-1-84943-758-5

Cover credit: 'The chasm is closed,' ©2005 Thomas Doyle.

Printed, bound and converted
by CPI Group (UK) Ltd, Croydon, CR0 4YY.

For Effie

Acknowledgements

Lungs received its first performance at the Studio Theatre, Washington D.C. USA on Wednesday 28th September 2011, directed by Aaron Posner in a rolling World Premiere with Paines Plough/Sheffield Theatres. The cast of the Studio Theatre production was:

Brooke Bloom

Ryan King

A Paines Plough and Sheffield Theatres production was first performed on 19th October 2011 at the Crucible Theatre, Sheffield, directed by Richard Wilson with the following cast:

Kate O'Flynn

Alistair Cope

The play was given a workshop reading at the Manhattan Theatre Club with support from the James Menzies-Kitchin Award and the Ian Richie Foundation, with Charlotte Parry and David Furr. The play also received a workshop reading at Paines Plough with Ruth Wilson and Geoffrey Streatfeild.

Thanks to Daniel Evans, David Muse, Aaron Posner, Richard Wilson, Adrien-Alice Hansel, Annie MacRae and all at MTC, Simon Stephens, Roxana Silbert, Pippa Hill, George Perrin, James Grieve, Tara Wilkinson and all at Paines Plough and Sheffield Theatres, Linda McLean, Mike Bartlett, Lyndsey Turner, Amy Rosenthal, Dan Rebellato, Clare Lizzimore, Nick Gill, Charlotte Westenra and Lucinda Burnett.

Thanks to David and Jean Heilman Grier and Jon and NoraLee Sedmak Thanks also to Jessica Amato, Rachel Taylor and Jessica Cooper.

Special thanks to Christina Pumariega.

This play is written to be performed on a bare stage. There is no scenery, no furniture, no props and no mime. There are no costume changes. Light and sound should not be used to indicate a change in time or place.

A forward slash mark (/) marks the point of interruption in overlapping dialogue.

A comma on a separate line (,) indicates a pause, a rest or a silence, the length of which should be determined by the context.

The absence of a full stop at the end of a line indicates a point of interruption, a trailing off or an interruption of thought.

There is no interval.

****

The play should be set in the city it's being performed in. Any references in the text that suggest another place should be amended.

The letters 'W' and 'M' are not character names. Any programme materials should simply list the actors and not who they are playing.

Lights up.

W A baby?

M Breathe.

W A baby?

M I was just thinking.

W About the future.

M We'd have to change how we live.

W The planet, use less

M no, that's, well yes but that's not

W okay.

M I'm freaking you out.

W Not / freaking me out.

M Completely. You thought you'd be the one.

W No.

M The one to say it, yes. To say yes, yes okay, I'm ready, yes, let's do it, yes.

W That's

M to put the pressure on, yes, / to try to convince me to

W pressure? Put the pressure on, I'm not a a a a

M we're having a conversation. That's all that's happening. All that's happening is we're having a conversation.

W You're having a conversation.

M We're having a conversation.

W A conversation you're starting.

M A conversation I'm, yes, that I'm trying to start.

W A conversation that you're deciding to start now.

M Yes.

W In Ikea.

M I hadn't planned to.

W No. Okay. Yes. Okay.

M Do you want some water / or

W that kid with the panda is staring.

M You're hyperventilating.

W Don't exaggerate.

M If it's too much

W it's not / too much.

M If it's too much we can put it back in the box, just put a lid on it and lock it away and then later when you're feeling less freaked out / we can

W I'm not freaked out.

M Alright fine okay.

W I'm not freaked out I'm just
surprised. I'm surprised I'm
fucking shocked actually. I'm

M freaked out.

W I'm not.

M You are.

W I'm completely freaked out yes because why don't you ever, how can you, why didn't you, why would you not talk to me about this / I wish you'd let me IN I wish you'd let me IN to your head. Into your fucking impenetrable fucking

M I'm talking to you now. I'm telling you now. We're talking, we're talking now, we're having a conversation. When should I have

W we're not. We're not. This isn't a conversation.

M Okay.

W It just isn't.

M Okay.

W I don't know what it is but I know for fucking certain it's not a

M right okay okay.

W Can we at least get out of the queue? Everybody's

M of course, I'm sorry, I didn't mean to just

,

W yes. I need a minute. Can we put it back in the box?

M There's no rush.

W Just to

M there's no hurry.

W Catch my breath.

M It's a conversation.

W Bit of a walk or something. Ten minutes. Meet you back at the car.

M Okay.

W What's wrong?

M You said ten minutes.

W I needed to think.

M It's pitch dark. You stink of fags.

W It's snowing. Is it snowing?

M You've not got a coat.

W This weather is insane.

M Coldest winter ever they've just said. Hottest summer, coldest winter.

W And you left the engine running.

M I was listening to the radio.

W I'm okay.

M I know I just

worried.

W No need.

M Good.

,

W Did we get any of the stuff we came here for?

M I went back but they'd

W shit.

M Yeah.

,

W A baby?

,

M I was just

thinking.

,

W Can we just

we will talk about it but

M I know.

W not right now. I'm too

M yeah, me too.

W Can I drive?

M Course.

W You can play your tape. Let me hear your new songs.

M They're not finished.

W Okay, well,

let's just sit and not say anything then okay?
Just be silent, just not have to deal with this right away
because

M good.

W I don't have the

M it's okay. Whenever you want to talk about it we / can

W no okay of course good but not now I don't have
anything to say about it right now because it's such
a shock, it's such an enormous, you can't just say
something like that to someone you can't just say that

to me and expect me to just be fine and rational and clear-headed / and not

M	when would be the right time to / mention

W	I don't know I don't have the answers I just know that *that*
wasn't
it.

,

I'm sorry.

M	I shouldn't have said anything.

W	No, no, you're right. You're right.
It is something we should

M	should we?

W	We should be, yes, be talking about, because, fuck, we're not getting

M	I know.

W	Any younger.

M	No.

,

So are we talking about it or

W	no.

,

Yes.

,

Go on.

M With what?

W With, you were saying, with, you know, what?
What were you saying?

M I've said it all.

W Then say it again because I couldn't hear you before because people were staring and I was pushing a trolley and holding a lamp and I couldn't breathe.

M You got the gist.

W I think so.

M Or you wouldn't have got as freaked out as you did.

W Touché.

,

Touché.

,

So.
Where do we go from here?

M Well, we should try to leave the car park.

W Sarcasm? Right now? You think that's going to

M I'd like to hear your opinion.

W Yes.

M Of course.

W Of course yes.

M It's a two-way

W I know.

M It's a two-way thing and so

W but

M go on.

W Alright,
it's *this* it's

I have no idea. I don't
opinion? I don't have
it's like you've punched me in the face then asked me
a maths question / while I'm still on the

M like I punched you in the face?

W You know what I mean.

M No.

W Okay. Yes. Let's do it. Let's do it. Yes. Let's do it. Yes.

M I'm not

W I'm saying yes.

M I'm not asking a question.

W Aren't you?

M No.

W You're starting a conversation.

M I'm not sure if

W well it's started and now it's happening and
I'm saying yes.

M Right.

W Look, alright, listen, you have to understand alright,
I'm thinking out loud here so please just let me talk

just let me think it through out loud please alright
don't just jump in if I say something wrong or stupid
just let me think okay because I've always wanted
alright and I'm talking in the abstract I've always
wanted I've always had a sense or an idea of myself
always defined myself okay as a person who would,
that my purpose in life that my function on this planet
would be to and not that I ever thought about it like
that it's only now because you're asking or not asking
but mentioning, starting the conversation only because
of that that I'm now even thinking about it but it's
always sort of been a given for me an assumption ever
since I was a little girl playing with dolls I mean long
long long before I met you, it's never been what I
guess it should be which is a a a a a an extension of
an expression of you know, fucking *love* or whatever,
a coming together of two people it's always been this
alright and this will sound stupid and naïve but it's
always been an image, I guess, of myself with a bump
and glowing in that motherly or pushing a pram or
a cot with a mobile above it or singing to it reading
Beatrix Potter or Dr Seuss, I don't care, never cared
about it being a boy or girl just small and soft and
adorable and with that milky head smell and the
tiny socks and giggles and yes *vomit* even it's all part
of it, looking after it, caring for it that's I think that's
the impulse and there's always been a father in the
picture but sort of a blurry background generic man,
I'm sorry, it's just this picture of my life I've always
had since I was able to think and I've never ever
questioned it. Never. And I've pushed it all down
and focussed on my career, on my studies, on myself
and now it's becoming, potentially becoming a bit
real I'm going to have to think about it for a second
please just a a a a a a or much longer in fact because
well because I'm not an idiot, I'm a thoughtful, very
thoughtful person and I want to do everything for the
right reason or at least a good reason and I believe in

questioning and never just blindly accepting or and it's going to take a lot of effort to unravel or to to to to to to excavate not excavate but excavate all of those previously held beliefs and assumptions because it's important probably the most important thing you could do to bring another person a yes a person an actually living breathing thinking because they won't stay small forever and I think don't they I think a lot of people think about them being small, just tiny and sweet and unconditional with their eyes and giggles and tiny little fingers gripping your thumb and I do I did I think I thought like that because it's too hard like we're not quite designed to be able to fully comprehend the the the the

M enormity

W or whatever which is a maybe it's a survival mechanism perhaps an inbuilt thing because, fuck, if you thought about it if you really properly thought about it before actually doing it then you'd never ever actually fucking do it because it's too fucking too fucking

M enormous

W it is it is it is it's fucking enormous, fucking enormous fucking the purpose of life itself, the purpose, the meaning, the meaninglessness, the love and the horror and the hope and the fear and everything the volume of all of it turned right up, the rest of your life the rest of someone else's life committing someone to something forever, ancestry, the seven and a half thousand generations of human history and I don't even know much about my own *grand*parents, let alone my *great* grandparents or or or or and it's *their* genes, their genetic stuff, really, them, these dead people moving around making choices for this little, this tiny, but that's not the whole thing,

M breathe.

W This is what I'm saying this is what I'm saying because they don't stay small, they grow up and become people, they become grown-ups like everyone else, they become their own grown-up people and they think their own thoughts and they buy their own clothes and they leave home and they hate you.

Alright, because I'm thinking out loud.

I'm thinking and talking.

M I didn't say anything.

W We're having a conversation.

M Yes.

W That's all that's happening.

M Look, let's get home and drink some gin and pretend I never said anything.

W And the planet.

M I don't want ice. Take my ice.

W The planet. Because you worry about the same things I do, you care about the same

M do I?

W And they say don't they that if you really care about the planet, if you really care about the future of mankind then don't have children.

M Do they?

W I mean, they actually say if you really care about the planet then kill yourself but I'm
I mean,

I'm not going to do that.
So,
because there's, what, there's seven billion people or so, there's too many people and there's not enough of everything so really the right thing to do, the ethical thing to do is to not contribute to that, particularly people like us.

M People like

W car driving, plastic bag using, aerosol spraying, avocado importing, Western,

M but we're good people.

W Exactly. We are we are we are good people yes we are. Good.

Are we?

M You can't think about that stuff.

W No I know. It's not our responsibility.
And anyway, so much about it is unknown.
And what if this kid, this hypothetical
what if she, or he, this imaginary little Edwin
or Hannah,

M Edwin?

W What if she or he was the person to work it all out and save everything, everyone, the world, polar bears, Bangladesh, everything, we don't know so

M no, but

W or we could plant a forest. We could work out the carbon footprint of the expanding nappies in the landfill and the Baby Gap hoodies flown in from the Congo or wherever and we could plant trees, entire forests, make something pure and and and oxygenating, so

M how do you factor that in?

W Exactly.

M The world is going to need good people in it.

With everything that's happening.
We can't just leave it to the people who don't think, the people who just have child after child without ever properly examining their their their their *capacity for love.*
I mean, that's what's wrong with everything isn't it?

W Yes. I know. Exactly.
Hang on what?
Are you saying some people are too stupid to have children?

M No. No of course not. But
yes.
Some people, lots of people, aren't thinking it through, not fully, and maybe the smartest, most caring, most informed people aren't having children.

W Right.

M So it's their genes that aren't surviving. So things are getting less caring and less informed and more savage.

W So, to save the planet it'll be, what, eugenics or

M no, no.

W Sterilise? Exterminate?

M Not what I'm saying.

W Camps? Enforced

M no. Of course not I'm not
I don't have the answers.

Yes, some people are saying that maybe that will happen but we'll be long dead by the time that's

I mean, you know more about this stuff than I do. You're the one doing the PhD.
But, yes, if we're being honest, really, teen mothers in tracksuits with fags in their mouths, smacking their kids in supermarkets, being a gran by thirty, multiplying like rats,

W rats?

M Meanwhile the people who read *books*, the people who *think* and try to help and I know I'm being a bit fascist here, I'm just playing devil's advocate here of course I am but there are some thoughtful people who are waiting for the perfect circumstances and there's no such thing as perfect so the world is overcrowded and people think well I don't want to bring my child into this world full of crack dealers and pimps and homeless, and I know this sounds reactionary but let's not be politically, you know, correct about this for a second, there are *some people* who just shouldn't have children. They just shouldn't.

And would it be such a great loss if those people, you know,
couldn't
have
children? Or
I mean,

isn't this what you were saying? I'm only carrying on from what you were saying.

W I think I'm going to be sick. No I'm not. I am actually I think yes.

M I was only

W I know that's what I feel, what I think sometimes, but when you say it out loud it sounds like the worst, cruellest, sickest, most hateful

M what do you want me to do?

W Just a cuddle and shut up for a bit I think would be good actually.

M Is that okay?

,

W We should adopt then maybe, probably, shouldn't we?

,

Why aren't you saying anything?

M Yeah, you're right, you're absolutely the best thing to do, absolutely, with the world as it is would be to

W so many unwanted, unloved

M and it's completely irrational, I know it is, but I

I don't
want
to do that.

I know that makes me sort of a

terrible person. I just don't think I'd be

W what if it's something I really want to do?

M Then we can talk about it.

W But not do it.

,

M Yeah.

,

W Right.

,

M I worry I might be one of those fathers who doesn't notice his kids unless they're winning stuff or getting in trouble.

W I don't want to be one of those mothers who only lives through their children. I want to still read books and do things. I will not use having a child as an excuse for becoming an idiot.

M You're not your mother.

W You're not your dad.

M I want to be able to play with my kids without it having to be competitive or educational.

W I want to still have sex. We mustn't let it ruin our

M people get so boring and it doesn't have to be like that.

W I don't want to have to host the best birthday parties or make the best Chewbacca costume for Halloween.

M Or push our kids to do stuff they don't want to do.

W Harp lessons or

M but it has to value learning and be able to think for itself.

W But not so thoughtful that it gets depressed and lonely.

M Autumn babies get picked first for sport.

W	No princesses or soldiers. No guns and tiaras. Disney will not dictate what our

M	and the schools are a mess, we'd have to get on the board of governors or parents associations.

W	We're talking about it. Look at us.

,

M	Are we too young to be thinking about this? To be worrying about all this? We used to do stuff. Go to the zoo. Go clubbing.

W	We still do.

M	When?

W	How about tonight?

M	Not tonight.

W	Friday.

M	It'll be too hectic on Friday.

W	Wednesday.

,

M	Okay.

W	I'VE MISSED THIS!

M	WHAT?

W	I SAID I'VE MISSED THIS. BEING OUT.

M	BEING OUT, YEAH.

W	WE'VE GOT SO BORING. STAYING IN. TELEVISION.

M	TELEVISION.

WHAT?

W IT'S VERY LOUD IN HERE.

M I CAN'T HEAR YOU.

W IT'S GREAT.

,

ARE YOU READY TO GO HOME?

M Fresh air.

W My feet still hurt. Used to love those boots.

M Look at those llamas.

W Everyone here has a pram.

M Shall I grow a beard?

W Yes okay yes it's yes let's make another person. I gave myself a week to think it through and yes I think we should do it. I think we should try.

M You're sure?

W Yes.
Aren't you?

,

You're not or you wouldn't be

M I'm sure. Yes.

W Okay.

M And you're sure.

,

W Yes.

M Okay.

W Let's go to bed.

M Okay.

W Can we stop a second?

M Again?

W I think I just need

M it's been weeks.

W I'm sorry I'm just

M not again.

W If we're / going to do this

M there's no *if* there's no
so don't please don't / start with that.

W I'm not starting I'm just
I'm talking, I'm saying that if

because I mean, not if, *because, because* we're doing this

M we are we are

W you're going to have to

M I am.

W Let me finish, you're going to have to relax because
this should be beautiful and

M I'm fine.

W You're fine okay good good because it's just, and
maybe I'm reading this all wrong maybe I'm seeing
something that's not really happening, but it feels like
you're sort of feeling quite a lot of

that the way you're being towards me is sort of
giving off a lot of

hate.

M I don't hate you.

W No I know I know you don't I know that I know that I know you don't.

M I don't hate you.

W Please don't take that the wrong way.

M Might go for a walk.

W I'm sorry I'm sorry I've ruined it I'm tense and it's not just your fault

M what isn't? Fault? What is?

W No, stop, okay look all I'm

okay. So.

,

We're trying. We're trying and it's
that's wonderful, it is and it's
scary it's wonderful and scary and not just for me
because it's my body it's going to be happening to,
not to but inside or whatever but for both of us it is
isn't it, and we're not talking about it and there's this
atmosphere isn't there or is it just me who's

,

maybe it's just me then.

M I was there. I was ready.

W Yes. Yes.

Yes I know.

I know and I'm sorry.

I know.

,

But it's

alright.

Deep breath.

I want our

it's about making a person. What we're doing.
It's about this amazing
miracle, not miracle but you know what I mean,
miracle yes miracle it's about this miracle happening
and it
I want it to
need it to feel
I don't know
sacred or
not sacred but
yes.

,

And you've got that porno look in your eyes.

M I can't help the way I look.

W okay okay okay and that's
in the right *context* that's fine, that's more than fine
that's

W That look that murderers and men in porn films have.

M It's just

I'm concentrating.

W Scares me.

M Really?

W No.
No of course it doesn't.
But yes, a bit.

M Fucking hell.

W Don't feel bad about it I don't want you to

M do you feel this a lot? When we

W no. No. No. No.

M You've said it before.

W Sometimes but

M you've mentioned it. It's one of your things.

W Only sometimes it
things? One of my things?

M I want you. Sometimes I get this
when I want you
get this animal fucking
horrible you know

lust
I suppose,
want to
fucking

you know.
Hard.
Want to
yes, hurt you maybe. A little. Make you scream.

W I know baby and sometimes that's really

M you want it.

W Sometimes I do I do sometimes I do but

M sometimes I want to get inside you so much, want to open you up, split you apart / like you're a

W okay okay okay and that's
in the right context that's fine, that's more than fine that's

but this is different, this is

I'm thinking about
aren't you thinking about

what it means? What we're doing? The thing beyond

the moment?

M Honestly?

W Yes honestly.

M Honestly I'm just thinking about the moment.

W Okay.

M In that moment I'm not thinking I'm just my cock and my mouth and my / hands and looking at your

W yes alright alright alright okay, good okay.

M So squeamish about this stuff.

W I'm not I'm just

just want a

I don't know. Connection. Is that what I mean?

Some

like it's the two of us together and not something you're

doing
to me.

Did you know in those scanning machines the same bit of a man's brain lights up when he looks at a woman as when he looks at a spanner?

M What does that mean?

W Sorry. Sorry.

Sometimes that look you do is sexy. It is. Sometimes it makes me think
oof.
And your shoulders and your noises and the weight of you and the danger, the

M danger?

W the way you need me, that there's nothing else on the entire planet for you at that moment, bombs could be going off, an earthquake or whatever and it wouldn't matter to you in that moment and I love I can make you feel that way and I do, often, usually, lots of the time I'm there as well, I'm absolutely
the world's not there it's just us and
we're the whole universe.

And then sometimes

and I'm just being honest here okay but sometimes it looks like you're about to hack off my limbs.
You know? Like you're going to smash in my teeth, throw me into bin bags and bury me in the woods.

Not always but that's how it feels sometimes.

M Fucking hell.

W Only sometimes. And only for a second.

M Fuck.

W I shouldn't have said anything.

M No.

W I should have just

M no it's

W shit.

,

I'm sorry I'm just scared and I'm not

,

I don't know what I'm talking about. I never feel that way.

Let's

,

woo. Fucking hell.

I need a big laugh or a big cry.

,

Know any good jokes?

,

We won't sleep now will we? Not for a bit.

,

We can try again. I'm glad we talked about it. We can try again.

M Yeah.
We will.

Not right away though.

W No. Okay.

,

I / love you.

M I'm going to read for a bit. Watch TV maybe.

W Do you want me to come with you?

M No.

Get some sleep.

W Okay.
Sweetheart,

just,

sorry but

don't

you know.

I know you're
that you didn't

M what?

W Doesn't matter.

M I wasn't going to masturbate if that's / what you're asking.

W All I'm saying is we need all the help we can get so please don't

M fuck sake.

W I'm just saying.

M Okay.

,

W Okay.

M Good morning.

W You're cheerful.

M I think spring's here.

W You made breakfast.

M We should move.
Go to Brighton. Isn't that what people do?
Get some outdoor space. Fresh air. Room for goalposts. Trampoline. Paddling pool. Trees.
We should plant some trees. Put a bit more oxygen into the world. Like you said. Do our bit.

W We should get married.

M Let's think about it.

W You don't want to?

M One thing at a time maybe.

W I'll make coffee.

M You should get a job.

W Once I've got my PhD my / prospects will be

M we can't both be

W I know this isn't very feminist of me but actually sweetheart I think you should, if we're serious about this, you should get something a bit more full-time and I know you get free records and time off to do gigs but we've both got to make sacrifices and I'm not going to be able to work once I get bigger and then for a lot of that first year and we're going to need to

M plenty of musicians have children.

W Plenty of successful musicians have children.

Sorry. I didn't mean that. It's just
the economy is in freefall and nobody's buying vinyl anymore. You really don't want to get married?

M What do you mean happening to? Last night when we were talking you said it's your body, that it's your body it's happening to.

W Well it is.

M Talk about it like it's a

W it is my body that
when it happens, yes, it'll happen to me.

M Alright I know yes of course but
do you want some of this?

W I'm not hungry.

M You make it sound like a
I don't know
damage,
violence,
an act of terrorism or

W well,

M you see it as

W yes.

M Threatening.

W Not just that but
actually I will have some
but, yes I'm fucking terrified. If I'm honest.

M That's

it's a shame.

W It's realistic. I'm trying to be

I'm readying myself.

Yes I'm excited about growing bigger and getting scans and yes giving birth as well of course but it's going to be painful and uncomfortable and my feelings, my thoughts are going to be all over the place.
And I've got anxiety about that. Of course I do.

You get that right?

,

I'll stretch, expand, become a a a a a house,
my breasts will swell, ache, hurt like hell then get drained and lose their shape forever, my

you know, my

I mean,
that's bound to change with what it has to go through.
Have you ever seen a real birth? It's not like TV. It's blood and shit and mess and I'll be torn and bruised like I've gone under a truck.

This must be how a caterpillar feels as it cocoons itself.

M I'm sorry I can't
that I can't do it.

W You don't have a womb.

M That's what I'm saying.

W You wish you could be the one who gets

M yes.

W Pregnant?

M Don't laugh at me.

W You wish you could gestate the foetus to full term then birth it through your

M a bit yes I do. I feel already that we're not equal somehow, that I can never quite know what it feels like and you'll know that I can't and why are you looking at me like that?

W You're sharing your anxieties.

M I'm sorry.

W No. I love it. This is delicious, is there more?

M Have mine.

W No.

M I'm done.

W What's wrong?

M Nothing.

,

You should quit smoking.

,

If you're serious / about

W yes you're right.

M You're going to be a home, an ecosystem and you're / polluting

W I've said yes shut up yes I've said yes alright so / please just

M it doesn't make sense to me that you could want to be a mother and still be smoking.

W You don't understand because you've never been addicted to anything.

M Those two impulses seem to me to be completely

W you're right.

So shut up.

,

Good luck.

M I'm doing it for you.

W For us.

M That's what I mean.

W You look handsome in your suit.

M I feel sick.

W How did it go?

M Find out Monday.

W Keep checking your email.

M Here goes.

W Well?

M You're looking at the newest cog in the corporate machine.

W I'm so proud of you.

M Will you drive me?

W How many do you think?

M Left here. How many

W trees. Few weeks ago you said about planting trees. Do you know? I do. How many trees would we have to plant to counteract the

M left again.

W I did some maths. How many plane trips? London to New York?

M I

W two thousand five hundred and fifty.

M two / thousand

W I could fly to New York and back every day for seven years and still not leave a carbon footprint as big as if I have a child.

M You're having second thoughts.

W Ten thousand tonnes of CO_2. That's the weight of the Eiffel Tower. I'd be giving birth to the Eiffel Tower.

M You / wouldn't be giving

W and if we had a second it doesn't just double because the chances of them reproducing and how many they might have and how many their children's children might have and how many their children's children's children might have that goes up exponentially. Fuck recycling or electric cars, fuck energy efficient fucking light bulbs, unless educated, thoughtful people like us stop making babies the world is totally fucking fucked.

M We should talk about this.

W We are talking about it, it's fucking nuts it's fucking terrifying it's the taboo, the last real taboo. / There is no incentive to save lives right now, there is no incentive to to to to to cure AIDS or whatever, to keep people alive, what we need is the planet to fucking *purge* us, fucking drown us, burn us, cull everyone by about two thirds.

M What's happening now? What's actually happening in your mind? What have you been reading? Because really it's important to check your sources thoroughly. There's a lot of scaremongering and I know you know that, I know you know all of that. You tell me off for the stuff I read because it's too too too I don't know.

W Fucking, hurricanes. Floods. Fucking volcanoes. Earthquakes. Tsunamis. Bring it on.

M Up here on the right.

W And we're just letting it happen. We're so caught up with our little fucking lives and we're killing everyone.

M We're not killing everyone.

We're not killing everyone.

,

W I got my period.

M Oh sweetheart.

W That doesn't mean I'm not right.

M I know.

W I'm going to smoke a cigarette and I don't need you judging me, alright? I'm going to smoke 'til I vomit.

M	Do you want me to take the day off?

W	Yes. No. Your first day? Don't be stupid.

M	I can.

W	We're here now. Go on. Just come straight home after.

M	Okay.

W	And bring cake.

M	Here.

W	What's that?

M	Cake.

W	I missed you.

M	Why? I mean, thanks, you too, but

W	lately it's feeling more and more odd when we're apart.

M	Sorry I've barely been here. My appraisal's coming up.

W	I wish I could come to work with you.

M	You could meet me for lunch. We could make it a regular thing. Tuesday thing.

W	Okay.

M	What's that?

W	Sandwiches. Picnic.

M	Shit.

W	It's Tuesday.

M	I forgot.

W Made us a picnic. This weather. Let's sit on the grass. Find some shade.

M Look, I can't.

W Didn't make it. Bought it. Got you a donut and a Ribena too.

M I forgot, I'm sorry.

W And an apple to counteract the donut. What do you mean forgot? It's Tuesday.

M There's a meeting.

W At lunch?

M Carbon efficiency, ethical blah blah blah.

W Skip it.

M I asked for it.

W Go in late. You / asked for it?

M Offsetting, that sort of thing yes I asked for it.
I think we should plant trees. The company. Lots of them. Forests of them. Try to offset our you know, our footprint, our

W yes.

M I've been reading that book of yours. Was up half the night. I've overtaken you with it I think. I've got to that bit about how since the Industrial Revolution we, people, everyone, we've put half a trillion tonnes of pure carbon into the air. Twenty seven billion tonnes a year.

W I had to stop reading it.

M Apparently an elephant weighs a tonne. So there's half a trillion elephants worth of carbon up there. And it'll only take us forty years to burn the same again. Our kid would be middle aged, maybe have kids of their own. Grandkids even.

It's actually pretty interesting when you get into it, when

W sex we should have sex. Find somewhere. Now. Walk and eat. Have donuts and if we find somewhere a a a a I don't know, a secluded

bush or
wooded area, or a
you know, a public
toilet,

M not a public

W we could be quick. You could make your meeting.

,

M What kind of donut?

W Iced.

,

M Alright.

W Well.

Well well well.

M We haven't done that in a while.

W Needed to.

M You're loud.

W Was I loud?

M Where did that come from?

W What does that mean?

M I'll have some explaining to do.

W There'll be other meetings.

M Yeah.

W Proud of you. It's good, it's a good thing that you're trying to do.

M I missed the meeting.

W Are we good people?

M Yes.

W I mean, yes I know but are we actually though?

M Yes we are.

W How?

M How?

W In what way are we?

M We just are.

W Yes. Okay.

M We're going to be great parents.

W I think it's okay to ask the question.

M So do I.

W Good.

M It's part of what makes us good people.

W But I don't
we don't believe, do we, in good and bad.
Right and wrong.

M Don't we?

W Don't believe in evil.

M Not evil no, we don't condemn people, we try to empathise, to put ourselves in their / position.

W Shoes I know we do yes but doesn't everybody think that they're good? Doesn't everyone believe they're

M some people wouldn't ask, wouldn't question it.

W Hitler or Rupert Murdoch or

M everyone thinks they're doing the right thing.
Pretty much.

W So what makes us sure?

M You're worrying too much.

W Yes.

M You're thinking too much.

W Okay.

M Okay?

W We must be very certain, arrogant even, to want to create another person out of our genes and to teach / it and to bring it up as

M we are we are we are certain. We're not
bad
people.

W Okay. Good.

M I'm going back to work now.

W Okay. I love you.

M It's going to be very lucky to have such thoughtful parents who care about things and who will love it very much.

W It?

M Him or her.

W We recycle.

M We do.

W We don't keep the tap running when we brush our teeth.

M That's right.

W We watch the news. We vote. We march.
We recycle.

M Yes.

W We support the smaller coffee shops against the larger chains.

M Even when it tastes like soil.

W Does it? Sorry.

M I'll see you tonight. I'll run you a bath and we can sit and talk. How's that?

W More hot.

M Okay. Move your feet I don't want to scald you.

W We watch documentaries. We read books about proper things. We read the classics. We watch subtitled films.

M We ride bikes. We buy fair trade.

W We give to charity.

,

What?

M I didn't say anything.

W We do. We give to charity. I do fun runs. I've done one anyway and I'll do another one.

M I didn't say anything.

W You have a red credit card for Bono's AIDS in Africa thing.

M I didn't say / anything.

W You gave five hundred pounds to the crisis appeal when that horrible thing happened. Okay too much hot now.

M Sorry.

W We shouldn't feel guilty about having things when really in the grand scheme of things we are not spoilt, we don't live beyond not too far beyond our means it's not swimming pools and sports cars and

M I'm not saying anything.

W We live pretty simply actually, we spend money on food and books and music and films and holidays sometimes and our mortgage and we don't just throw it away and yes it would be lovely to give more I do feel like we should, I wish we could give more but it's just not, we don't have any more

M hey, just
alright, you're

W am I being mental?

M You're somewhere on the spectrum.

W Oh fuck it's maybe it's hormones maybe it's my hormones I do feel crazy I do feel like some, I don't know, some chemicals some foreign chemicals are

M alright still, you're still

W sorry I'll stop I'll stop I'll just sit for a second I'll just try to calm down, just soak for a second do you think that's what it is? Do you think the garage is still open do you think they'll, they sell condoms and things don't they but a pregnancy test do you think they'll

M do you want me to go and see?

W Don't ask please don't ask that why are you having to ask me isn't it obvious, isn't it
sometimes you look at me like I'm a cryptic fucking crossword. Do I have to say it? Yes, yes, yes go, check, go and check now why are you not putting your shoes on why are you just staring at me like I'm a

M is this still the hormones do you think or are you just being nasty now?

W I'm sorry oh fuck I'm losing my mind this is amazing does this happen? Is this what happens or am I just having a breakdown? I fucking better be pregnant because otherwise this is terrifying.

M Isn't it vomiting? Isn't that the first thing?
Or missing a period?

W It's been keeping track of
what date is it?

M It's

W thirteenth, fourteenth

M when did you

W cinema when did we go and see that film?

M You mean the

W that shit film that you liked and I fell asleep in?

M That was

W was that three, four weeks ago already?

M You didn't like that film?

W Five. I think it was fuck I think yes it was fuck fuck
fuck it was I should have already had
go. Go. Go. Why are you still here? I'll come with you.
I need a towel. Give me a kiss.

M How long does it say?

W A minute. Two minutes.

M Alright.

W Let's leave it for three just to make sure.

M Check it each minute but then keep checking.

W Good good yes good. I'll come and get you.

M Right okay.
What?

W I can't go when you're right there.

M You've peed in front / of me before, you've

W I know but I've got performance anxiety as it is and

M some water I could get you some

W please yes but then will that count if it's come / straight
through you

M or will it have to sort of percolate I see what you're
saying.

W It doesn't say in the instructions.

M Do I really have to leave?

W This isn't one of the special bits. Pissing on a plastic stick isn't one of the sacramental

M yes. I want to, I want to see all of it I think it's all

W fine okay okay okay run the tap or

M how's that?

W Ssshhh.

M Gush or a trickle?

W Please.

M Sorry.

,

Shouldn't keep it running like this.

W Can we not think of the planet for one second?

M Sorry.

W Okay that's
I think
yes,

,

right.

Now we wait.

,

How long's that been?

M Thirty seconds.

W Really? Wow.

M A minute.

W How about

M two minutes.

W Right.

M Three.

W Okay,

here we go.

,

M Where's my phone?

W You hate your parents.

M That's not true.

W Okay, I hate your parents but
I mean they're not like mine, mine are

M you could have called them already.

W I'm going to dial. Do you want to speak to them?

M No.

W After, I mean, they'll want to speak to you.

M Why?

W I don't know, they might, they might want to say, you know

M I could call mine at the same time.

W You could of course you could, just let me, I just, let's just get this first call done and then, you know, then people know, then it's out and

M you want them to be the first to know.

W Is that okay?

M Before mine.

W I just

yes.

M Okay.

W I'm tingly. What is it I'm feeling? Like, naughty a little bit, like

M you shouldn't feel naughty.

W Like embarrassed almost, like I'm

M it's a secret that's why.

W It is it is it's our secret. It's our secret. Fucking hell I just had this fucking massive, ooooaar, shiver up my spine.

M You've gone white.

W Fucking goose bumps I'm shaking look I just had this and this is going to seem so weird but I just had this thought which was that it's not just you and me any more, in this room, it's you and me and then there's this other thing here now which is half you and half me and half completely something new.

It's happening. It's happening. It's happening.

M Yes.

W You look normal or is it just that I'm not seeing clearly?

M No, I feel
I'm happy. I'm ecstatic.

W Good. Good.

M It's not sunk in yet I don't think.

W No. I know. I know. What do you mean it's not sunk in?

M Nothing, it's

I've been preparing myself for it, bracing myself and now

W shock.

M Exactly.

W But you are
you're feeling something of course you are, you're not just

M I'm
yes I'm
it's the eye of the hurricane maybe or

W no, I know I know.

M I'm watching *you.*

W Right.

M Getting everything from *you*, you're

W buzzing.

M Should we call your parents?

W Okay.

M No, I mean, should we?

W Because

M they do say don't
not straight away
because it's still very early and a lot can

W why are you saying this?

M Just, they say it for a reason because, you know, they say one in four pregnancies ends in

W can we for just one second feel invincible and reckless before we start saying words like miscarriage?

M I just think

W do we have to be thinking right now? Can't we just live inside this perfect little bubble of happiness for a moment?

You're stood so far away

,

M Yeah, this is
this is all a bit
I'm just a bit

W ignore me, that was
you don't have to feel or react any particular way or you're right you're right, you are. I'm just
really

I'm not smoking as well so

cranky. First day of school feeling. The world just got bigger and I can see us from space and we're just a

speck.
I'm going to call my mum.

M Do you really hate my parents?

W Yes.

,

M Because

why?

W The thought of our offspring sharing their genetic code. That I could look into my son's eyes and see your father makes me

no. I'm being emotional. Ignore me. They're fine. I'm fond of them in a way. Of course I am.

M We never stay. Not long.

W That's because
don't you feel this
that in their own way they're lovely, they're
once you get attuned to their
you know,

but if I stay there for any length of time I worry that I might just put some scissors though their necks.

M Right.

W Don't you feel that way?

M What, about your parents?

W No, about

why would you feel that way about *my* parents?

M I don't.

W Why would you say that? My parents have never been / anything but

M they have I know. I'm not

W you should be careful.

M Sorry.

W And our kid is going to have their genes too so be careful.

M Can we just

never mind.

W What?

M Never mind.

W What?

M Can we
I need to just put this on pause, just put this argument / on pause and say I love you.

W Not an argument it's a / conversation.

M if we're going to be tearing bits off each other it's good that we take a second just to

W can you not do this?

M we're parents now.

,

W Yes. Okay.

M So do you want to hug or maybe

W no let's save that for

M no, okay.

W I don't feel like doing that right now.

M Fine.

,

Let's call your mum and dad.

,

W Maybe later.

M Yeah.

W Okay.

M So what did she say?

W Oh, you know.
She's happy. Of course she is.

M Except

W nothing. She's pleased, she's happy.

M They didn't want to speak to me?

W No.

M Congratulate me?

W She was crying.

M Good crying?

W Crying for, I don't know. I don't think she means what she said.

M What did she say?

W She knows I love you.

M What did she say?

W She's just anxious for me that's all.

M Wishes you were having someone else's baby.

W I didn't say that.

M Well you're not, you're having mine so she can go fuck herself.

W Please don't / talk about

M bitch bitch bitch bitch.

W No.

M Thinks I'm boring that's what you said.

W No.

M Because I told her I was reading a book on radiation.

W No.

M Because I tried to talk to her about food scarcity and economic incentives for capping emissions.

W Now you *are* being boring.

M They were *your books.*

W She didn't say anything.

M Give me the phone.

W Okay.

M I'm going to call her.

W Fine.

M I will.

W Go ahead.

,

M I'm going to prove her wrong.

W I know you will.

,

I know you will.

Come here.

,

What's the matter?

M I can't sleep.

W Try.

M I've been trying for hours.

W Switch your brain off.

M It's just disaster scenarios playing out in my head.

W Oh baby.

M Keep drifting off but it's explosions and helicopters and chunks of land collapsing into the sea.

W Anxiety dreams.

M No shit.

W You've seen too many crap films.

M We should at least / consider

W I know but just

M when then, when? We need to be equipped for if they

W I'm not going to know until they say. I can't plan for how I'm going to feel in that moment if they're saying our child is

M okay but

W that there's something wrong with

M We should have thought about it before. Talked about it.

W Okay alright here it is.

I want it whatever. If it's
I don't know

I'm not comfortable with flushing it away and starting again until we get one that's perfect.

M I'm not saying that but let's be / realistic about

W I'm saying I know what I want.
I don't care if it's born without bones or limbs, legs or a face just a a a a a an amorphous blob of skin and a mouth which just screams and screams I don't care if it's like the thing in Eraserhead it's mine, ours and it's part of us and we'll love it, we will, we'll love it no matter

no matter.

,

M No. I know.

W Won't we?

,

M Yes.

,

Let's try to sleep.

,

Are you asleep?

,

I can't sleep.

I'm trying.

,

I've always thought

I'm okay, I'm an okay person. A normal enough person. Basically, you know, good.

,

I wish I'd read more when I was younger. Or just, maybe slept with more people. Or travelled. It's going to be harder to travel now you know, with

,

nothing's ending. It's the start of something.
I'm young. We're young. It's exciting.

Can't tell if my eyes are open. Am I keeping you awake?

Just can't stop thinking. Everything. Life. The universe.

Death.

Try to picture what I'll look like. Dead. Stopped. Just another object in the room. Who'll see me like that.
What my expression will be.
When the face just relaxes it looks sort of sarcastic I think.
My dead face.

If it's going to happen, the solution to it all, the survival of mankind, it will happen in our lifetime. It has to.
And we'll be alive to witness it.

When we're talking I can't take my eyes off you.
You're so honest with me and it hurts sometimes but I remind myself you only say these things because you trust me. That I'm an anchor or something, I'm a nucleus to your proton, is that right? I think sometimes you think we're having a conversation and that I'm listening and responding but really I'm only silent

because I have no idea what to say. You're like an animal, a hungry, wild animal and you're circling me and snarling and you're beautiful and exciting and you're trusting me because I'm standing still but really I'm just frozen to the spot, really I just don't know what else to do and I worry that sometimes you think that that's strength and it's not, it's not, it's just

I dreamt that the baby had been born but it was just a cloud. A little thunder cloud. Or a squirming little creature with fangs and claws and flashing eyes.

I'll take a book off the shelf, any book and you'll have turned down the pages and underlined things. Put stars in the margins. And I stare at the sentence you've highlighted. I'll reread the paragraph. I'll think what has she seen that I can't see? What is it that I don't understand?

W Go to sleep.

M What time is it?

W You've been snoring.

M What are you doing?

W Vomiting. Feel poisoned. Burst a blood vessel in my eye, look.

M Oh. Yeah.

W Look, my stomach. I'm definitely

M no, you are I can see it.

W I'm

M massive.

W Oh.
I wasn't thinking

M no I mean,

W massive? Really?

M Just, no, showing, you're

W I just meant

M I mean you can see it, you can, you're definitely

W you can see it, you can definitely

M there's a bump.

W A bump. I've got a bump.

,

According to the Internet, my uterus is the size of a grapefruit.

Why do they use food to explain everything?
And why do they call the baby 'baby'. Not 'your baby'?

'Baby is the size of a lentil.'
'Baby is the size of a kidney bean.'
'A peanut.'

'An olive.'

M Are you hungry?

W Famished.

M I'll make you breakfast.

W I want bacon and washing powder.

M You can have bacon.

W Close your eyes.

M What have you done?

W We have a nursery.

M Green?

W We don't want pink or blue and yellow is so

M you're covered.

W I've painted little trees.

M I brought you chilli chocolate.

W You're my hero.

M Is it crazy I've been looking into local schools?

W Do you want to know? Boy or girl? Or should / we

M should it be a surprise on the day?

W Not that the day won't be special enough as it is.

M I'm too curious to wait.

W Me too.

,

I was thinking.
It's silly.

What if I don't
what if I can't

love it?

What if I look in its eyes and I let it grip my little finger and I hold it in my arms and I don't feel a thing? That happens doesn't it?

M Not to us.

W I'm so terrified. I have no idea how I'm going to be.

M You're going to be a wonderful mum.

W You don't know that.

M Yes I do.

,

W Thank you for being my person. For putting up with me when I'm

,

we should get married. I know that's old fashioned or whatever. I know you're against the idea.

I love you. I don't say that enough.

M You look beautiful.

W Not fat?

M Fat and beautiful.

,

W Going to get bigger.

M I expect so.

W Much.

M Yes.

W I'll be a planet.

So
you know.
Brace yourself.

,

M Okay.

W Ouch.

M What is it?

W Oh.

Oh no.

M Blood.

W Not this. Not to us.

M Hospital. Now.

W I can't bear this.

M I'll wait outside.

W No.
I need you.

M Okay.

,

W At least I can start smoking again.

M Oh sweetheart.
Shit.

W You were right. We shouldn't have got excited.

M Don't be silly.

W We shouldn't have told anyone. We knew this could happen.

Do you have change for the coffee machine?

M Give me the keys.

W I'm not an invalid.

M What do you want me to do?

,

W It's for the best probably isn't it? Would have completely taken over our lives. So much time and money, so much that could go wrong. End up sleepless with worry or with our heads in the sand.The world. Who'd want to have a child now? We should be happy. This is a relief it is it is it is we should take a deep breath and give a huge sigh of thank fuck for that because we're not going to add to any of it, we're not going to add yet one more lost person into this crowded little world so good for us. Let's crack open some champagne. Let's fly somewhere. Spend our money on *us* before the global economy completely implodes. When the riots start let's join in. Let's smash something. Start some fires. If we see an electric car with a baby on board sticker let's ram it off the road the fucking hypocrites. The damage they're doing.

,

It's a relief for you I suppose.

,

M Why would you say that?

W Just

you never quite seemed as

M that's not true.

W It wasn't as necessary or

M it was my idea.

,

W Yeah. I'm just trying to

I'm just trying

,

M get some sleep.

,

I'm here.

I heard you get up, can I sit with you?

You fell asleep. I brought you in here.

Goodnight.

Good morning.

Goodnight.

I made some tea.

I'll see you after work.

Have you been sat there all day?

It's getting dark.

It's getting light.

Come to bed.

You fell asleep.

I woke up and you were gone.

You've not said a word to me for days.

,

Baby.

,

We need to get on with things.

W Please don't.

M It feels like you're punishing me.

W I'm not trying to.

M You're angry with me.

W No. No I'm not I'm

yes. Alright I must be I suppose. I'm angry at this.

This everything. I hate it.

M We can try again.

W No.

I can't bear it.

Sorry.

There it is.

,

Should we stay together do you think? Or should we

I don't know.
Not.

,

M I

W I'm not saying this because I want to break up.
I don't know what I want.

I just think it would be
I think it's a good opportunity for us to
you know,
talk.

Have a conversation.
Ask ourselves some serious
you know,
difficult

M opportunity?

W You know what I'm saying please fucking help me out here.

M No. I know what I want.

W We should, no, we should take some time and really think about this because right now I don't want to look at you. I need you to just fucking put your arms around me but you're not and don't, please, not now, not because I'm telling you to, I don't want that, not

M sorry.

W Fucking apologising you don't know do you what you're apologising for, so

I need some
just
space.

M Space?

W Space. Fucking space. Yes. The fucking final frontier why are you / repeating things I'm saying?

M Sorry I'm
just
getting my head around it that's all. Just trying to catch up a bit, trying to fucking

W get your head around it.

M Yes. And it's taking me a second so I'm sorry. I love you. I'm

I don't know what you need or what you
this is difficult for me too.

W Oh fuck off / I'm not saying it's not, I'm not saying it's
easy for you I know it's
don't put that on me don't
look we're both in the same
we've both

M just I need some help I need some help a little fucking
I need you to give me some clues here as to what
you need because honestly I feel like you're standing
behind a glass wall just this sheet of glass and I can't
reach you.

W Don't cry for fuck sake because

look,

for fuck sake.

,

It's okay.
Fuck. It is.
There are people in the world with real problems.
This is just

people go through this everywhere. All the time.
Through time.
We'll
we could try again. If that's what we want to do.

M It is.

W I don't know.

,

Feel like I've had my skin peeled off.

M I kissed someone else. At work.

The new girl. The temp. I didn't talk to her about you and I don't know why, didn't say I had a

,

she was nice to me. She was nice to me. We talked about music.

And she kissed me and I didn't stop her.

However hurt you feel and raw and angry that's how I'm feeling too and that I love you more than anything. I'd cut off my arms for you. I'd pull my eyes out. And I'm sorry. And I know that if I go back in on Monday and she's there I might do it again.

I could see our future together, this picture of what our lives were going to be like and it feels like I've been burgled and I have to live in this stupid angry version of the world where the person I adore, my best friend in the world can't even look at me and where I can't look at myself.

I miss you.

I want to sit opposite you in a restaurant, share a bottle of wine. I want to go to the cinema and hold your hand. Run around in the park.

We can try again.

Adopt maybe.

,

Why are you smiling?

W For the first time in ages I feel a little clarity.

I'm sad. I've been really sad and I don't know if that's ever going to go away completely. I'm sorry I've not

been able to tell you what I've needed.
I don't know what I needed.

Yes I do.

I needed you to be patient with me. To wait as long as it took. I needed you to be braver than me and put your own feelings second and to understand, even when you didn't understand. To use your initiative for once and not need instructions. To try to imagine what it's like to miscarry. To realise that there are certain feelings I'm trying to cope with and protecting you from and I'm working *hard.*

Most of all I needed you to not kiss someone else.

,

So, I'm being sensible for both of us. That's it.

Lucky we don't have kids.

,

So there you are.

M You look good.

W No I don't, shut up. Thank you.

M No, it's

W thanks for meeting me.

M Oh, it's
no, of course. I was surprised when you asked

W right.

M Starbucks.

W Oh. Yeah. Well. You know.

M Didn't seem like you.

W Beard.

M I know.

W It's good.

M Oh. Yeah.

W I didn't know who else to tell.

M No, I'm glad you did.

W Mum was very fond of you.

M I thought she hated me.

W Yeah.

,

M How was the funeral?

W It's tomorrow.

M Oh.

W Come with me. Will you come with me? I mean, if / you want to.

M Really? Because

W weird? Too weird, forget it. Sorry.

M Of course, I'll, yeah if you want me to.

W I do yeah that would be good. You don't need to check with

M check with what? What?

W Nothing.

,

M My dad died. Few months back.

W You never said.

M No, I know I
didn't know if

W oh, no, that's

,

We're getting to that age now.

Scary. Be just us left soon.

,

Fuck. Look at you.

M You too.

W Just want to just

take just a second.

Mental picture.

Click.

M It's alright.

W No, I know. I know.

M Oh, please don't
you don't need to
oh.

W I'm sorry, I promised myself I wouldn't

M it's okay. It's okay.

W Mess. Already. Brilliant. Well done me.

M You're okay, just

you're okay.

,

W How are you anyway? How's life? How's work? Good? Everything good?

M Yeah / it's

W how's the temp?

M Oh. Yeah. No, we're not

W no?

M No we're, we, / I mean we

W that's really I'm really I'm sorry about that.

M No no no, I mean after we,
you and I
once we
I just wasn't
I didn't have the
energy
for another

not right away, not

W so you didn't
I mean
with the

temp

that didn't
you didn't

M oh, well, yeah but
I mean, that was just

fucking hell.

That was this whole other

W right.

M Thing.

W Right. Yeah.

M How about you?

W Do we really want to talk about this? I mean,
I don't want to talk about this.
I know that we're not
that you and I are no longer
but I still don't want to hear about your your your
sex life or

M no okay.

W So you're not together now?

M No. Fuck. No. No.
No.

No.

W Right. Okay.

M Now I'm with

W oh okay.

M Yeah.

W You don't have to

M I don't mind.

W I don't want to hear it.

,

So okay.

This is horrible.

I don't know what I was expecting.

I think I might leave before we start talking about the weather.

M It's fucked.

W It is. It's impossible. Sweat's pouring off me like

M everything's shut down. City can't cope with it.

W Keep passing out.

M What?

W Dehydration or
I carry a bottle around with me but I just go through it like

M yeah.

W So I take two but then I'm just lugging around all this water and it's

M I know I'm the same it's

W we weren't right together were we?

We were good people, weren't we, but
it just wasn't meant to be.

M You don't believe in any of that predetermination, destiny stuff.

W I do now.

M Oh,
sorry. Really?

W Yes.
No.

I think what with the world how it is,
I think if I didn't think someone was going to fix it all,
that some superhuman genius was going to work out
how to fix it all then

I think I'd just entirely lose my mind.

I don't know.

I'm different.

We both are.

I quit smoking.

M Yeah?

W It's shit. I feel so much better.

M I started.

W You're kidding.

M I look cool.

W No.

No.

M How about your course?

W Oh. Yeah. Finished. Done.

M You're a doctor.

W Technically.

,

M I've been playing again. Couple of gigs in a friend's band. Nothing big.

W Yeah?

M It's a sort of electronic thing, a new direction.

W Right.

M I think I've got stupider.
Since we

,

I don't know what books to read, I don't really read anything.

I have no idea what's happening in the world.

W It's fucked.

M I can't remember how to think.

And it feels, you know, really

wonderful.

Not having to worry about stuff I have no control over. I used to get so angry at people who didn't read or think or care about anything but I completely get it now.

Sometimes I'll be driving. No, it's stupid.

,

Sometimes I'll have this thing, I'll just sort of
like I wake up.
While I'm driving. Not that, exactly, because I'm not tired, all I do is sleep. It's not that. It's just

who's been driving? I don't remember anything from the last twenty minutes. You know. Like I've been on autopilot. Zombie or

and I have *days* like that.

Days.

,

Now I'm here with you and

,

W yeah.

,

M Yeah.

W Well.
Well well well.

Fuck.

M Yeah.

W Well.

M Yeah.

W Well well well.

M That was

W it was. It really was.

M I don't feel bad about this.

W You,
what? You don't

M not at all.

,

W Oh.

M What?

W Nothing.

M What?

W No, nothing I'm

I was being
I was thinking something.

M What?

W Just

letting myself get carried away for a bit.
I forgot for a second. About her.

M Oh. No, I didn't mean

W it's fine.

M I
no I'm
I wasn't thinking, I was just

W no of course.

M Just talking just

W it's just
it felt right.
Didn't it? I mean
to me anyway.

M Yeah, no, of course it was
yes.
It did.

,

It did.

But

W can we save the buts? I think we both know.

Can we just
for a minute
just lie here and pretend that we're years ago and
nothing's happened. That we'll get the papers
and maybe go to the pub for a roast. Play the quiz
machine. Have a quarrel about some book I've
been reading that I explain badly and sound like I'm
advocating

I don't know,
compulsory euthanasia of the old or

you know, I kept a drawer of your socks and things.

It's stuffy in here. Sorry. Can't breathe. I'd open the
window but it's just as bad out there.

I've not slept with anyone else since you.

You don't need to know that. It doesn't matter. It
doesn't matter that you slept with two people.

,

What, more?

M We were apart for

W no I know. I know.
It doesn't matter. How many? It doesn't matter.

M You want to know?

W Yes. No. I've not so much as kissed anyone else.
I'm crying again. Shit.

I tried to. I went for a drink with this guy. But he was
horrible.

,

M Don't tell me her name. Your girlfriend.

,

M Fiancée.

,

W Oh.

M Yeah.

,

W Let's never see each other again. Okay?

,

You look startled.

M You said we shouldn't see each other.

W Yet here I am. Why might that be?

M It's raining.

W Do I look different?

M You've not got a coat. Different?

W I thought I might, yes
look

different.

Glowing maybe.

,

M You mean

W yep.

,

So, okay, that's as far as I planned. I got as far as telling you then I hoped you'd maybe jump in and save the day, be all manly and know what happens next because this is
I mean,
come on Superman. Swoop down. Save the day.

Anything.

,

Nothing.

I've broken you. You've shut down. It's too much. Can I come in at least? I'm drenched.

,

It'll be born by the time you've said something. It'll plop out of me and be slippering, squirming around in the afterbirth can I at least come in or

M listen,

W it speaks.

M Okay, listen, I'm sorry but can I

can we talk later? I'll come round, I'll

W she's here isn't she?

Of course she is. I'm an idiot. This is horrible. I'm going.

M Wait, just

W I can't do this.

M Please.

W Don't.

M I knew I'd find you here.

W How?

M I looked everywhere else.

W You know me so well.

M Shouldn't be out here at night.

W We had sex in those toilets do you remember?

M I'm sorry I couldn't speak before.

W I made you miss your meeting.

M Obviously I'm going to do I want to do the right thing.

W The right thing. Good.

M I'm going to support you.

W Support. Great.

M Will you just please, I'm sorry, but just be quiet because it's you're making it impossible to think and I really need to think.

W No. I won't. You shouldn't need to think. You shouldn't. You should just fucking know what to do. Say what you want. Say what you're feeling. Don't tell me what I need to hear.

M I don't know what you need to hear.

W Then it shouldn't be a problem.

M What I'm feeling, what I'm thinking is there is no perfect outcome of this. I have to tell my fiancée.

W Yes you do.

M And I
please, shut up for a second.
I'm sorry, just give me a chance here.

,

Thank you.

,

Right.

So
okay,

let's get married.

W No.

M Why?

W Okay, one, because you're already engaged to someone else.
Two,
because that's the least romantic proposal in the entire sad sorry two hundred thousand year history of the human race and three,
because I don't want to. Fucking hold your horses I don't even know whether I'm going to keep it yet so let's not get ahead of ourselves.

M You're not / going to

W I've not decided yet and don't tell me your opinion because it's fucking irrelevant.

What's your opinion?

M All I'm saying is

W you think I should abort it.

I wish you were dead.
I wish I was dead. I need an earthquake. A tsunami.

M I'll do what it takes. I'll do whatever you want, whatever you need.

W I need you to call your fiancée.

M I'm going to talk to her. Now. Face to face.

Okay? I'm doing it.

,

Can I come to yours once I've seen her?

W Why?

M Because I want to.
And I'll have nowhere else to go.

,

W Okay.

M Okay.

W You're bleeding.

M She was upset.

W And there was

M kicking. Screaming. Hitting.

W Did she hurt you?

M Scratching. She caught me on, under my ear, her ring

W good.

M Yeah,
I suppose.

W	No she deserved to. She deserved to really fucking

M	yeah.

W	Cripple you.

M	No, you're right.

W	For what you did.

M	Yeah, but

W	should have cut your fucking dick off.

M	Okay.

W	I must be insane. You clearly can't be trusted.

M	I

W	you're clearly unable to overrule your prick.

M	I'm not entirely to blame for

W	yes actually yes fucking yes you fucking are, yes.

M	It was both of us who

W	I'm not engaged to her.

M	Nor am I now.

W	Touché.

,

Touché.

,

M	So what now? We just
what? Go back to normal?

W	Normal? What normal? What?

M No, just

W we're not a couple.

M Oh.

W We're not
we're not going to be like

M no, right.

W Not like before. We need to start fresh. If at all.

M I just
sorry, I thought

W I don't know you. You're a fucking stranger now.
A fucking deceitful, immature evil little stranger.

M I just assumed.

W Well don't. This is all just going at a hundred miles an hour and I need a second to breathe.

M No. If we think about it too much we won't do it.

W Do what?

M This. Now. Us.

Yes it's not the perfect circumstances, but let's go into this with open arms. I love you. Okay? I always have. When I'm away from you I forget how to enjoy anything and when I'm with you I feel at home.

We've never worked out how to be together without making each other feel a bit shit and I want to find a way to not do that. You've got to stop ripping bits off me and I've got to grow up and behave like an actual human being.

You've needed me to know what you need without having to ask.

You've needed me to be aware of how I'm feeling and to let you in to my head.
Right now I know exactly what you need to hear and it's absolutely what I'm feeling.

We're not going to overthink this.

We're doing this.

We're going to get the books and go to classes and work out how to be parents. And we're going to grow old together and look back on all this and laugh because it will seem like a different lifetime.

And we'll have a conversation and we'll just try to do the right thing. Because we're good people. Right?

And we'll plant forests. I mean it. We'll cycle everywhere. We'll grow our own food if we have to. We'll never take another plane. We'll just stay right here. And we'll plant forests. You and me.
You, me and the little speck. The ten thousand tonnes of CO2 waiting to be unleashed onto the planet.

,

W Thank you.

I needed to hear that.

,

I'll think about it.

,

M Breathe in.

W Okay.

M Breathe.

W They're getting more frequent.

M Every few minutes.

W Get the keys.

M We're here.

W Don't leave me.

M I'm right here.

W I'm scared.

M You're so brave.

W Hurts.

M I forgot the camera.

W Here it comes.

M He's so beautiful.

W Let me hold him.

M Of course.

W We're a family.

M I'm so tired.

W It's your turn.

M Shall I bring him in with us?

W He's got to wean.

M I know.

W The books all say so.

M Good marks all round.

W I'm so proud of him.

M I forgot my camera.

W There he goes.

M Don't cry.

W He's never slept away from home.

M We're going to be fine.

W You shouldn't let him see that.

M I'll talk to him.

W He needs to hear it from his father.

M Are we going up this weekend?

W I just want to see him before he's off because, what with the world as it is,

M I know.

W Have you got your camera?

M Let's get married.

W I do.

M We need to be prepared.

W It won't be as bad as they're saying.

M He hasn't called.

W They've suspended all flights.

M It's so hot.

W The planet's fucked.

M He looks so different.

W He wants to help out.

M We're fine.

W That's what I told him.

M It's a standard operation.

W I know.

M Just goes with old age.

W He's going to drive you.

M I'm going to be okay.

W I'm scared.

M You're fine.

W He's found me a home. They have classes and things there. Art and things. It's nearer to him so a long way from here. He gets cross with me sometimes. You know how he is. I think a lot of people are angry at me. At us. Those of us still around. I forget more and more. I don't know what they're so upset about.

I miss talking to you. Here I am talking to myself.
Your forests have gone. I don't watch the news any more, it all just gets worse and worse.
Everything's covered in ash. He tells me to stop dusting. Snaps at me.

I'm tired. Fed up.
But I'm okay. Listen to me moaning on. It's a nice cool day today, like we used to have. Fresh air. No sirens.
No noise. Nothing.
It's good.

,

Anyway, I just thought I'd stop by. Change the flowers.

I don't know if I'll get much of a chance to pop back here so anyway.

,

I love you.

,

Lights out.

by the same author

The Most Humane Way to Kill a Lobster
9781840025590

Monster
9781840027594

Don Juan Comes Back from the War
Ödön von Horváth, in a version by Duncan Macmillan
9781849432542

1984
George Orwell, in a new adaptation created by
Duncan Macmillan and Robert Icke
9781783190614

Every Brilliant Thing
9781783191437

People, Places and Things
9781783199099